R15

SAFARI PARK

Library Edition Published 1990

© Cherrytree Press Ltd 1989
© Marshall Cavendish Limited 1990

Library Edition produced by DPM Services Limited

All rights reserved. No part of this book may be reproduced or utilized in any form or by any means electronic or mechanical including photocopying, recording, or by any information storage and retrieval system, without permission from the copyright holders.

Printed in Italy

Library of Congress Cataloging-in-Publication Data

Graham, Alison.
  Safari Park / by Alison Graham
    p. cm. – (Lets go to)
  Includes index.
  Summary: Describes a safari park and the many animals that live in it.
  ISBN 1-85435-243-1
  1. Open -air zoos – Juvenile literature. [1. Open-air zoos. 2.Zoos.] I. Title. II. Series: Graham, Alison. Lets go to.
  QL78,088 1990
  590'.74'4 – dc20   89-17337
                                                                                   CIP
                                                                                     AC

Let's Go To a
SAFARI PARK

By Janine Amos
Illustrated by John Rignell

MARSHALL CAVENDISH
New York · London · Toronto · Sydney

Have you ever seen a lion? Have you ever seen an elephant? Have you ever seen a kangaroo?

You can see animals in zoos. You can also see them in wildlife parks and safari parks.

Let's go to a safari park and see the animals there.

kangaroos

At the park, we meet the chief curator. He is in charge of the park. He has a staff of curators and keepers. The curators are experts who know all about the animals. The keepers care for the animals and keep them safe.

Big animals like zebras and rhinos run free in fields called enclosures. There are fences or ditches around them. They keep the animals and people apart.

Some of the animals are tame enough to be allowed outside the enclosure. Here comes an elephant!

These animals are red pandas. They come from China. Once, there were lots of pandas. They lived in forests in the mountains. But people cut down the trees. They needed firewood and land for farms. The pandas had fewer places to live. There was not enough food for them all. Fewer and fewer baby pandas were born.

Some of the pandas were taken from the wild and brought to the safari park. Here, they can have all the food and space they need. Everyone hopes that, one day, they will have baby pandas.

It is the job of the safari park to save and protect wild animals. This work is called conservation. It is very important.

red pandas

tiger cub

Many of the animals in the park were not born in the wild. They were born here. The curator is very proud of them. He shows us the nursery for animal babies.

Most of the animals can look after their own babies. But some of the babies need a helping hand. In the wild, this tiger cub would eat food that its mother caught. Here, it has a plate of fresh meat to eat.

baby monkey

Helping animals to breed is important work. Parks and zoos try to breed animals that are in danger in the wild. If the animals breed well, the young ones can be sent to other zoos and parks. In time, they may breed, too. One day, some of the animals may even be sent back to live in the wild.

otters

The curator explains how each enclosure is designed to suit the animals that live in it. Some animals need trees to climb, or water to swim in. Some need shade from the sun. Some need to be in the open.

In the wild, otters live near water. So, in the park, they have a pond to play in. They love sliding in with a splash.

The animals in the park have come from all over the world. Some are used to hot weather. Some are used to the cold.

Elephants and zebras come from sunny Africa. They learn to go out into the wind and rain. But there are shelters for them if it is too cold.

meerkats

These animals are meerkats. They are from Africa. They live in burrows in the ground, like rabbits. They are the same color as the sandy ground they live in. When they stand still, their enemies cannot see them against their background.

Meerkats love the sun. The curator has given them a special heat lamp. On cloudy days, the meerkats lie under it, just like human sunbathers.

Chinese pheasants

These birds are pheasants from China. The gardeners at the park have made a special enclosure for them. In it, they have put plants that grow wild in the birds' natural home.

Other animals have plants from their home countries. The kangaroos have gum trees from Australia in their enclosure.

The animals like to eat the food they would have in the wild. The staff spends much of each day preparing special food.

Some of the animals have meals like ours. They have fish, meat, eggs, vegetables, and fresh fruit. Other animals have more unusual food, like insects, earthworms, and rats.

Some animals eat only once or twice a day. Others have lots of small meals.
The curator tells us that we must never feed the animals. Candy and potato chips make some wild animals very ill.

How the animals are given their food is important. Giraffes have long necks. In the wild, they pull leaves from the tops of the trees. In the park, they like to have their feeding racks placed up high. Then, they can feed as they would in the wild.

Lions and other animals that hunt for their food have to make do with meat which has been killed by someone else.

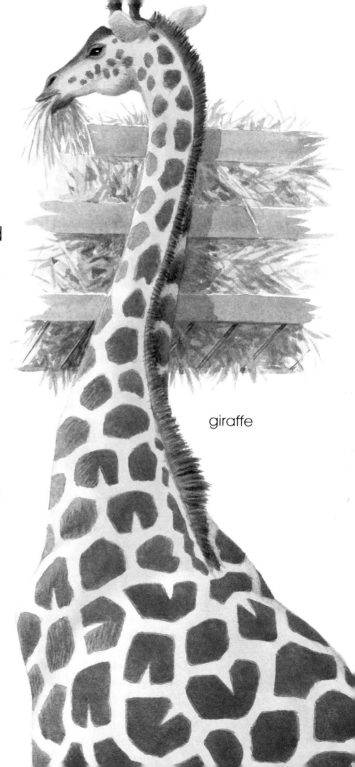

giraffe

penguins

Some of the animals have to be fed by hand. Penguins dive into the sea to catch fish in the wild. They will not pick up fish from the ground. At the safari park, the keeper throws fish to them. They catch the fish in their beaks and swallow them whole.

Watching the penguins is fun. They waddle along in lines by the edge of the pool. Then, they dive gracefully into the water. They swim around underwater as if they were fish, not birds.

keeper

flamingos

These birds are flamingos. Their pink color comes from the food they eat in the wild in Africa. At the safari park, the keepers give the birds medicines that contain the coloring. It keeps the feathers pink.

In the wild, the birds can fly away. In the park, their wings are clipped. They can still fly, but not very far. They cannot fly out of the park.

The keepers keep all the enclosures clean. They also keep the animals clean. Often, the animals like to play. The keeper hoses down the little elephant. Its mother gives the keeper a shower in return.

gibbons

Monkeys and apes love to play. In the wild, these gibbons swing through the trees with their long arms. At the safari park, they swing up and down their enclosure. They look like the human children playing on the swings outside.

The gibbons need to play to keep happy and healthy.

One of the gibbons' favorite games is eating peanuts and spitting the shells at the people watching. Gibbons have long delicate fingers. They peel nuts and fruit very carefully.

They like to watch the people who are watching them. They squeal and point and laugh at them.

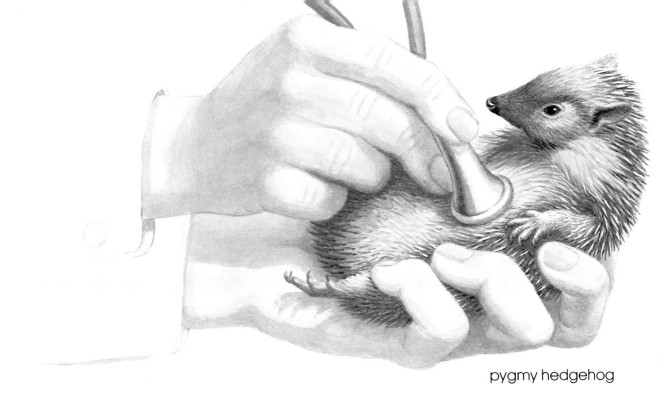

pygmy hedgehog

Sometimes, the animals get ill. Each day the keepers inspect all the animals. They notice if an animal looks tired or sickly. They keep a diary of each animal's looks and behavior.

When the vet comes each week, the staff tell her which animals need to be examined. Even the smallest animals need regular check-ups. The vet is listening to this pygmy hedgehog's heart.

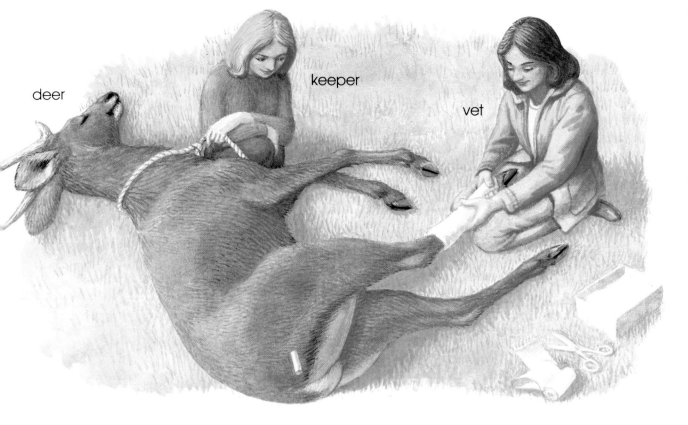

If an animal seems very ill, the keeper calls the vet immediately. This red deer broke its leg. It was in great pain, and it was very frightened. The keeper had to keep it calm until the vet arrived. She gave it a shot to send it to sleep.

Now, she can examine the leg and set it in plaster. When the deer wakes up, it will be able to walk without pain.

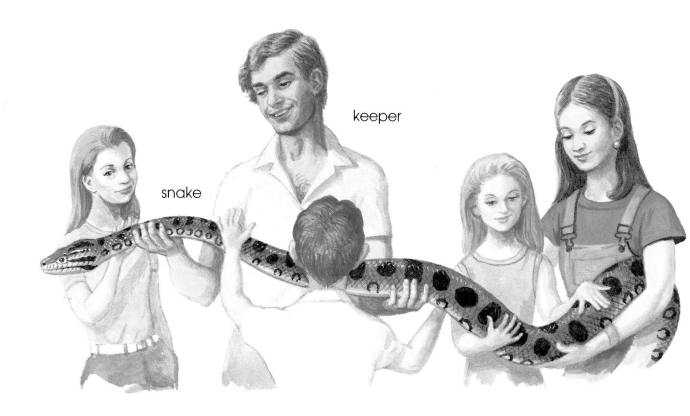

Now, we are going to see some animals close up. One of the curators shows us a huge snake. Some people are scared to touch it. They think it will be cold and slimy. But it is not. It is smooth and warm. Seeing the snake in real life is much more interesting than seeing it on the television.

At the children's farmyard, there are animals we can pet. There are tame goats, sheep, rabbits, and pigs. There are lots of ducks and geese.

We can take turns to have a ride on a pony.

We say goodbye to the curator and get back into the car. Now, we are having a special treat. With the windows firmly closed, we drive out of the park through the lions' enclosure. The lions can come right up to the car. They cannot smell us, so we are safe. They look at us, and we look at them. It is just like being in a real game reserve in Africa.

At the end of our visit, the curator takes us to the bird garden. He shows us the aviaries. These are huge cages made of strong, thin netting. Inside each aviary, there are brightly colored birds. They can fly around, but the netting stops them from flying away. It also keeps ordinary birds from flying in.

Fact File

Pandas like to play.

What is a safari park?
A safari park is a kind of zoo. Animals are kept in zoos so that people can study them without having to go to faraway places.

Once, most zoos were in towns. They were small, and the animals had to live in small cages. Now, people try to keep animals in large wildlife parks in the countryside. There, the animals can live almost as they do in the wild. In safari parks, there are dangerous animals like lions and rhinos, as well as other animals. They live in huge enclosures. People can drive through the enclosures in their cars, just as they would if they were on a safari in Africa.

Is it cruel to keep animals in safari parks?
Many people think that animals should not be kept in cages in zoos and parks. Other people

think that it is all right if the animals have plenty of space and do not get bored. They think that if people see how wonderful the animals are, they will want to treat them well in the wild.

How can safari parks help wild animals?

Some animals are dying out in the wild. People have turned the wild places where they live into farms. They have built towns and factories in the places – habitats – where the animals live. Safari parks raise animals that are in danger of dying out. They return some of them to the wild, so that they can breed there.

Safari parks also let people know what is happening to animals in the wild.

Who works in a safari park?

The people who care for each type of animal are called curators. One curator is in charge of reptiles. Another is in charge of birds. Another is in charge of mammals.

The curators and their research assistants are all scientists. They study the animals and find out new things about them. They watch them eat, play, and sleep. The same people often observe the animals in the wild.

The keepers look after the daily needs of the animals. They make sure they are well fed, clean, healthy, and happy. They get to know the animals very well and play with them if they are bored.

How do you become a keeper or a curator?

You have to learn as much as you can about animals and nature. You may be able to join a club at your local zoo or wildlife park. You may be allowed to help out with the animals. If you do well at school, you may get a job in a zoo and train to become a keeper. To become a curator, you need to study zoology at college.

What happens when animals are ill?

Sick animals are treated by the vet. They may have to have drugs or an operation just as people do. If an animal cannot be helped, the vet may have to kill it.

Index

Africa 13, 14, 20, 29, 30
apes 22
Australia 15
aviaries 28

baby animals 8, 10, 11
birds 15, 18, 20, 28, 31
breeding 11, 31
burrows 14

candy 17
children's farmyard 27
China 8, 15
cleaning 21, 31
conservation 8
curators 6, 10, 12, 17, 26, 28, 29, 31

deer 4, 25
ditches 7
ducks 27

earthworms 16
eggs 16
elephants 4, 5, 7, 13, 21
enclosures 7, 12, 15, 21, 30

feeding 10, 16, 17, 18, 23, 31
fences 7
fish 16, 18
flamingos 20
flying 20, 28

food 8, 10, 16, 17, 18, 20, 23
fruit 16

gardeners 15
geese 27
gibbons 22
giraffes 17
goats 27

habitats 31
heat lamp 14

illness 17, 24, 25
insects 16

kangaroos 5, 15
keepers 6, 7, 18, 19, 20, 21, 25, 31
kookaburra 28

lions 5, 17, 28, 30

mammals 31
meals 10, 16, 17
meat 10, 16, 17
meerkats 14
monkey 11, 22

nursery 10

otters 12, 13

pandas 8, 9, 30
parrots 28
peanuts 23
penguins 18, 19
pheasants 15
pigs 27

playing 12, 21, 22, 30, 31
pond 12
pony 27
potato chips 17
pygmy hedgehog 24

rabbits 14, 27
rats 16
red deer 25
red pandas 8, 9
reptiles 31
research assistants 31
rhinos 7, 30

safari park 5, 8, 18, 20, 22, 30, 31
sheep 27
snake 26
sun 12, 13, 14

television 26
tiger cub 10
toucans 28

vegetables 16
vet 24, 25
weather 13
wild animals 31

zebras 7
zoos 5, 11, 30, 31